# Amazing Nature

# Super Survivors

## Tim Knight

Heinemann LIBRARY

**www.heinemann.co.uk/library**
Visit our website to find out more information about **Heinemann Library** books.

To order:
☎ Phone 44 (0) 1865 888066
🖹 Send a fax to 44 (0) 1865 314091
💻 Visit the Heinemann Bookshop at www.heinemann.co.uk/library to browse our catalogue and order online.

First published in Great Britain by Heinemann Library, Halley Court, Jordan Hill, Oxford OX2 8EJ, part of Harcourt Education. Heinemann is a registered trademark of Harcourt Education Ltd.

Editorial: Jilly Attwood and Claire Throp
Design: David Poole and Geoff Ward
Picture Research: Peter Morris
Production: Séverine Ribierre

Originated by Ambassador Litho Ltd
Printed and bound in China, by
South China Printing Company

ISBN 0 431 16663 3
07 06 05 04 03
10 9 8 7 6 5 4 3 2 1

**British Library Cataloguing in Publication Data**
Knight, Tim
Super Survivors - (Amazing Nature)
591.5
A full catalogue record for this book is available from the British Library.

**Acknowledgements**
The publishers would like to thank the following for permission to reproduce photographs:
Ardea pp. **15** (Jean-Paul Ferrero), **26** (John Cancalosi); Corbis pp. **7** (Tim Davis), **14** (Yann Arthurs-Bertrand), **19** (Ecoscene/Joel Creed); FLPA pp. **8**, **13** (Wendy Dennis), **9** b (Jurgen & Christine Sohns), **10** (Minden Pictures), **20** (L. Lee Rue); NHPA pp. **5** (B. & C Alexander), **9** t (Nick Garbutt), **11** (Andy Rouse), **12** (Martin Harvey), **16**, **24** (Kevin Schafer), **17**, **27** (Stephen Dalton), **18** (B. Jones & M. Shimlock), **21** (Agence Nature), **22** (A.N.T.), **23** (Daniel Heuchin), **25** (G. I. Bernard); Tim Knight p. **6**.

Cover photograph of a lizard, reproduced with permission of NHPA/Martin Harvey.

Every effort has been made to contact copyright holders of any material reproduced in this book. Any omissions will be rectified in subsequent printings if notice is given to the publishers.

# Contents

Any words appearing in the text in bold, **like this**, are explained in the Glossary.

# make yourself at home

Plants and animals need certain things to survive. Their lives may depend on finding food, water, oxygen, sunlight, warmth, shade or shelter. Plants and animals have managed to make their homes in almost every part of the planet. Some, like tiny **bacteria**, can survive almost anywhere. Others have become specialists. They have found ways to live in conditions that most creatures could not stand.

## Adaptation

By changing how it behaves or looks, an animal or plant can become better suited to the place where it lives. These changes are known as **adaptation**. Adaptation happens very slowly, over thousands of years. Snow leopards in the cold Himalayan Mountains have thicker coats than other leopards to help keep them warm. Plants lose water through their leaves, so desert cacti have spines instead. The **surface area** of the spines is so small that hardly any water can **evaporate**. The African lungfish can survive out of water for months by breathing air. Although it has **gills** like other fish, it also has a pair of simple lungs. Polar bears are able to close their nostrils while they swim underwater.

By adapting to life in a harsh **environment** – such as the hottest, coldest, wettest, driest, darkest, deepest and highest places in the world – animals and plants can gain a big advantage. They are unlikely to meet many enemies and there is usually less competition for living space. They may also have a plentiful supply of food all to themselves.

4

Polar bears live in the frozen Arctic. Their coat of long, oily fur is almost waterproof. They even have hair on the soles of their feet, which protects their skin and gives them a good grip on the ice.

# Special diet

All plants and animals need food, wherever they live. Different plants and animals have different ways of finding food. Some survive by eating unusual food, or feeding in places that other creatures cannot reach.

Sundews and pitcher plants grow where most ordinary plants cannot survive. They live in areas with poor soil, such as bogs and rain-drenched mountain slopes. They add to their diet by feeding on a few juicy insects. Sundews catch insects in the sticky hairs that cover their leaves. The leaf tip of a pitcher plant grows into a vase-shaped trap. The trap is filled with a liquid that turns insects into plant food.

*Once an insect falls into the vase of a pitcher plant, the slippery, waxy walls of the trap stop it crawling out again.*

## Under pressure

Sperm whales feed on giant squid that live over a kilometre below the ocean surface. Deep in the ocean, the water is pitch black and ice-cold. The **pressure** of the water at such depths is strong enough to crush the life out of most animals. A sperm whale's rib cage is designed to fold up and collapse. As the whale dives, its lungs slowly shrink, like a balloon going down. Any air left in its body is squeezed into a tiny space. Hunting in total darkness, it can stay **submerged** for an hour or more without taking a single breath.

Flamingos gather in enormous flocks on Africa's salt lakes. They feed on tiny plants called algae. Algae are the only plants that can survive in such salty water. To collect its food, the bird lowers its head into the water and walks forward. It uses its upside-down, backward-facing beak as a sieve. The muddy bottom is so **caustic** that it burns the skin of other animals, but a flamingo's legs and feet are protected by a scaly covering.

*The pink colour of a flamingo's feathers comes from the algae on which the birds feed.*

# We all need water

Water is important for all **organisms**. In some parts of the world water is scarce. Many of the animals that live in these places must travel great distances in search of water. Others learn to survive without water for long periods of time, or make do with tiny amounts. Plants have to wait for water to come to them. When rain arrives, they need to store as much as possible.

Some desert animals, such as rats and mice, do not need to drink water. They survive on the small amount of moisture in the leaves or seeds that they eat.

In the Namib Desert, when fog rolls in from the sea, darkling beetles climb to the top of a sand dune. They put their heads down and their bottoms up, and wait. Tiny drops of moisture begin to collect on their bodies and trickle down into their open mouths.

*The male sand grouse stands in the water and uses his fluffy breast feathers as a sponge. Once they are soaking wet, he flies back to the nest. The thirsty chicks suck his feathers.*

When rain finally falls in the desert, cactus plants make the most of it. The saguaro cactus can take in about 900 litres of water in a single day. Its coat of spines has a much smaller **surface area** than ordinary leaves. When the hot, dry desert wind blows, the cactus loses far less water than a normal plant.

*Some saguaros can grow almost 15 metres tall.*

## Plant links

The American saguaro cactus and the African baobab tree are both built like water tanks. By soaking up rainwater from one heavy storm and storing it in their fat trunks, they can survive the long drought that follows.

*The baobab is nicknamed the 'upside down tree' because its branches look like roots.*

# Keeping warm

Cold can kill. To survive in freezing temperatures, animals must find ways of keeping warm. Whales and seals have blubber. This is a thick layer of fat under their skin. Blubber stops their body heat from escaping even in the coldest seas.

Emperor penguins are the only animals that can survive all year round in Antarctica. The males spend the winter standing on the frozen ice while they **incubate** their eggs. Their long, thin, windproof feathers protect them from freezing cold blizzards, which can reach up to 180 kilometres per hour.

Life in the mountains can be bitterly cold too. The yak is a wild ox found in the Himalayas. Its long, shaggy coat reaches almost to the ground. The vicuna is a member of the camel family that lives high in the Andes Mountains. It has a thick, fine woolly fleece that helps to keep it warm.

*Male emperor penguins huddle together to keep warm, taking turns to face the driving wind.*

*During the winter, macaque monkeys in Japan escape the cold by taking a bath in the hot springs around their home.*

### Green blanket

Plants that grow in cold areas also have ways to keep warm. One Himalayan plant, the saussurea, covers itself in a kind of woolly blanket. The Arctic willow tree avoids the wind by growing no more than a few centimetres above the ground. Cushion plants in the mountains of Tasmania cover the ground like moss. One square metre may contain up to 100,000 plants. By growing so tightly packed together, they hold in the warmth and avoid drying out.

# Staying cool

Overheating is just as dangerous as freezing. Vicunas would soon overheat if their thick fleece covered them completely. Instead, they have bare patches on the inside of their legs. They cool down by exposing this part of their body to the chilly air.

*Shovel-snouted lizards keep cool by balancing on two feet at a time and lifting their other two off the hot sand.*

Deserts are hottest in the middle of the day. Even in the shade, the temperature in the Sahara can reach 58 °C. Most animals avoid this burning heat by sheltering from the sun. They look for food in the early morning and late evening, when it is coolest. Others move around at night when the temperature is lower still.

Some animals do stay out in the heat of the day. They have other ways to keep cool. The Kalahari ground squirrel uses its tail as a **parasol**. It turns its back on the sun and curls its long, bushy tail over its own body. Moving within its own patch of shade, it can continue searching for food even when the sun is blazing down.

Sweating and panting help animals to lower their body temperature. Many desert animals, such as the fennec fox, jack rabbit and bandicoot, have huge ears that seem far too big for their bodies. These ears are not just for hearing. They are filled with tiny **blood vessels** that lie just under the surface of the skin. The bigger the ears, the more air blows over them. This helps to cool the blood as it flows round the animal's body.

A Kalahari ground squirrel enjoys a meal while sitting in the shadow of its own tail.

# Extreme heat

Although very high temperatures will kill most animals, some **organisms** thrive in them. Near the Galapagos Islands there are giant worms that are over 3 metres long. The worms live on the seabed, close to underwater volcanoes that spout boiling hot water. They feed on the huge numbers of **bacteria** that live and breed in the volcanic water. The worms do not have mouths. Instead, they take in food through their feathery **tentacles**.

*In the desert there is often less than 250 millimetres of rain in a year, so water is scarce. The fats that are stored in a camel's hump are enough to help it survive for about three weeks without water.*

Living organisms can also survive in the boiling hot springs of Yellowstone National Park in North America. Green algae are found in the bottom of the steaming pools. These simple plants grow together in a kind of green slime. Parts of the algae grow above the surface of the boiling water. These areas are just cool enough to allow flies to settle on the algae. Brine flies lay their eggs on the algae. When the **grubs** hatch, they eat the green slime.

## Fireproof trees

Certain plants can withstand the heat from a raging fire. Pine trees and other conifers thrive on it. The flames kill competing trees, but the bark of a pine tree does not burn. The fire does not harm the growing parts inside the trunk. The tree's delicate buds are protected by thick needles. These needles burn at low temperatures, so the flames do not last long. The fire passes quickly through the branches and moves on before its heat can damage the tree's buds.

Some plants cannot survive without fire. Many banksia trees hold their seeds in airtight capsules, which only open when they are scorched by hot flames. Then the seeds fall on ground that no longer has other plants growing on it and can quickly **germinate**. The layer of ash left behind by the fire also helps to **fertilize** the soil. Proteas growing in the South African Cape are even more dependent on fire. Their seeds will not germinate unless they are soaked in rainwater mixed with ash from a fire.

Some trees only drop their seeds after a bushfire has burnt away all the other plants.

15

# Breathing difficulties

No animal can live without oxygen. In places where oxygen is hard to come by, survival is difficult. All around the world, animals have different ways of overcoming their breathing problems. They have found ways to reach oxygen, to store it, to use less of it, or even to do without it for a long time.

Water spiders live, eat, **mate** and lay their eggs in an underwater bubble. They collect air bubbles from the water's surface and fasten them together to make a sealed chamber. Once they have finished building their see-through home, they climb inside.

*The Weddell seal uses its sharp teeth to cut breathing holes in the ice when the Antarctic seas freeze over in winter.*

*The water scorpion hunts by hanging upside down underwater, using its sharp claws to grab tadpoles. It breathes through the tip of its long tail, which sticks out of the water and acts like a snorkel.*

### Deep breathing

Whales spend their whole life in water. Even so, they still need to breathe air. When humans breathe out, we only release 15 per cent of the waste air in our lungs. Most whales almost completely empty their lungs with one massive, spouting belch from their **blowhole**. This allows them to refill their lungs with completely fresh air. Whales can also store extra oxygen in their muscles. As a result, they can go far longer than humans between breaths. They often dive for up to an hour.

Water lilies often grow in muddy swamps where there is very little oxygen. They help their **submerged** roots to breathe by pumping air down to them. But oxygen alone is not enough for plants. They also need a good supply of light.

# Sun seekers

Whether plants take root in water or on land, they cannot survive without sunlight. Giant aroids are found in the Borneo rainforest. They grow on the dark forest floor where there is very little sunlight. The giant aroids have massive leaves, known by the locals as 'elephants' ears'. These help the plant to take in as much sunlight as possible.

*Coral reefs only grow where the sunlight can reach them. Although corals are animals, they also have algae growing inside them, which survive by absorbing sunlight.*

## Light-gathering lilies

The giant Amazon water lily collects light through its floating leaves. Each leaf starts as a giant spiky bud. In just a few days, it can grow almost two metres across. Its upturned rim helps it to push aside other floating plants. One Amazon water lily can produce up to 50 leaves in a season. This is enough to cover large stretches of water and keep the sunlight to itself.

Plants take in sunlight through their leaves, so anything that covers them is bad news. The leaves of many rainforest plants end in a sharp point known as a 'drip-tip'. Rainwater runs down special grooves in the leaves towards this downward-pointing tip. Water drains away quickly, allowing the leaves to dry. This stops moss and other moisture-loving plants from growing on the leaves and blocking out the light.

*The Amazon water lily survives by making sure it gets most of any available sunlight. It also has spikes on the underside of its leaves so that fish cannot feed on them.*

# In the dark

Even in the parts of the planet that sunlight cannot reach, animals manage to survive in permanent darkness. They do this either by living without light or creating their own light.

## Cave life

Oil-birds are found in Venezuela and Trinidad. They spend the night searching for fruit in the forest. During the day, they **roost** in a cave. Like bats, oil-birds can find their way in total darkness. As they fly around inside the cave, they make clicking noises. The sound bounces back off nearby objects, such as rock walls or other birds. This tells them how close they are. The sooner the echo reaches them, the closer the object. This system is called echolocation and helps the oil-birds to avoid crashing into things in the dark.

The **antennae** of the cave cricket are extra long and super-sensitive. It uses them to find food in the dark.

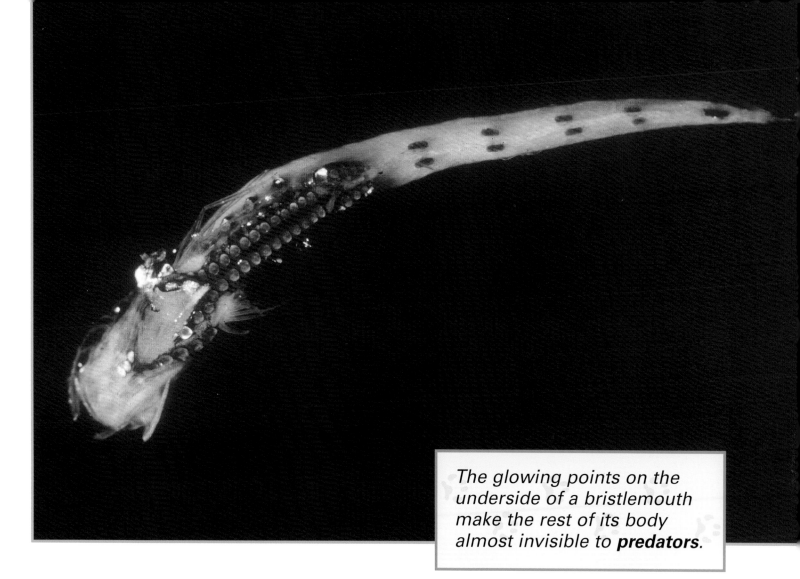

*The glowing points on the underside of a bristlemouth make the rest of its body almost invisible to **predators**.*

## The deep

The darkest depths of the ocean are pitch black and freezing cold. Many of the creatures down here are 'bioluminescent'. In other words, they glow in the dark. The light is produced by special chemicals inside the bodies of fish and other deep-sea creatures.

Scientists are only just beginning to understand why bioluminescence is so important, but they know that some creatures use it for hunting. In the deep, dark ocean food is hard to find. Some angler-fish use a flashing light to attract smaller fish from a long distance away. This light is at the tip of their long **dorsal** fin, which they use like a fishing rod. They bend the fin forward over their head and dangle the flashing light just above their open jaws.

# Leading a double life

Some animals and plants are able to survive sudden changes in their **environment**. Some of these changes, such as a rainy season or a dry season, happen only once or twice a year. Others, such as the rising and falling tide on the seashore, happen every day.

*During a **drought**, the Australian lungfish can survive in a small puddle of water.*

In water, the lungfish uses its **gills** to breathe, but it can also survive by breathing air. The swamps where it lives often dry up completely. When this happens, the lungfish tunnels down into the mud and buries itself in a hole. The lungfish breathes by sucking air into a pair of pouches that work like simple lungs.

The South American capybara, the world's largest **rodent**, spends the dry season grazing on the open plains. When the land is flooded during the rainy season, capybaras use their webbed feet to swim through the floods. They feed on water plants and grasses that most animals cannot reach. A capybara's eyes, ears and nostrils are all on top of its head. This allows it to keep most of its body underwater, hidden from **predators**.

Mangrove trees grow in the mud where the river meets the sea, and have to survive in both salt water and fresh water. There is no oxygen deep in the mud, so mangrove roots grow in a kind of platform. They spread their roots across the surface and take in **minerals** brought in on the tide. They take in oxygen directly from the air through spongy patches on their roots. Every rising tide covers them in salt water. Some have roots that **filter** out the salt. Others **absorb** it, and **expel** it quickly through their leaves.

*When the tide is high, many small fish escape from predators by sheltering among the mangrove's submerged roots.*

# A place called home

When competition for space is very strong, it is even more important to choose an unusual habitat (home) or a unique way of life. A rainforest is filled with vast numbers of weird and wonderful plants and animals. Every nook and cranny is turned into a home or a shelter. Those most likely to survive are the ones that make the best use of their surroundings.

Tent-making bats **roost** underneath a large banana leaf. The leaf acts like an umbrella that protects them from both sun and rain. They use their sharp teeth to cut a groove down the underside of the leaf, so that it hangs down like the sloping roof of a tent.

*Shelter from the storm. Tent-making bats stay dry by clinging to the underside of their leafy hiding place.*

An army ant nest may contain up to 150,000 individual insects.

### Living house

Army ants camp in 'bivouacs', temporary shelters built out of their own bodies. The outside walls are made up of soldier ants. They link legs and face outwards with their jaws wide open. Safe inside, the smaller workers turn themselves into dividing walls so that the living tent has separate rooms. These rooms contain the queen, her young and her servants.

Three-quarters of all rainforest animals never leave the **canopy**. A tree frog may spend its whole life in and around a single bromeliad plant growing high in the treetops. Some frogs even lay their eggs in the tiny pool of rainwater trapped by the bromeliad's cup of leaves, using it as a pond in the sky.

# New homes

Some animals and plants have adapted to living side by side with humans and are perfectly at home in towns and cities.

In North American gardens, racoons raid the dustbins, while hummingbirds sip sugary drinks from bird feeders. In Britain, foxes and squirrels are regular garden visitors. Cockatoos, parrots and possums hang around Australian houses in search of an easy meal. In tropical countries, house geckos hunt insects attracted to the bright lights. Japanese crows have even learned how to crack walnuts by placing them under the wheels of cars when they stop at traffic lights!

*A hungry racoon will find plenty to eat in a pile of rubbish.*

## House guests

**Roosting** bats have swapped their caves for dark attics. Huge flocks of starlings blacken the sky as they gather for the night in Britain's warm city centres. White storks nest on European chimney pots. Peregrine falcons dive down on pigeons from New York skyscrapers. In Brazil, thousands of purple martins leave the rainforest every night to roost inside an oil refinery. In spring, they fly north and nest in the man-made boxes put out for them by American bird lovers. Black rats, which used to live in the trees in Asia, now use their climbing skills to scale the walls of buildings. The house mouse adapted to living alongside humans many centuries ago. Wherever people make their home, they soon find themselves sharing it with a family of mice!

All these animals have taken advantage of new opportunities and made themselves at home in a different place. They are the latest winners in a long line of super survivors.

*The rooftops of European buildings, like this one in Spain, make perfect nesting sites for storks.*

27

# Fact file

The world's smallest tree is the dwarf willow, which grows to about 5 centimetres tall on the tundra of Greenland.

The Australian saltwater crocodile can survive in both fresh and salt water. It has been found up to 1350 kilometres out to sea.

Marine iguanas in the Galapagos Islands dive underwater and eat seaweed.

The Weddell seal can travel underwater for up to 10 kilometres without coming up for air.

A desert bighorn sheep can survive for a week without water. When it finally drinks, it takes in up to a fifth of its own weight in water.

The yapok, or water opossum, carries its babies in an airtight, waterproof pouch when it swims underwater.

Snowy owls keep their feet warm with feathered toes.

The fennec fox has hair on the bottom of its feet. This is to protect its feet from the hot sand.

Emus can hear the faint sound of raindrops falling on the ground, and follow this to find water.

A hibernating woodchuck's heart beats four times a minute, compared to 80 times a minute when it is active.

Bears, dormice and hedgehogs **hibernate**, spending up to seven months in a deep sleep and living off their fat reserves.

The Australian walking fish not only survives without water, but can also climb trees to feed on insects.

Hibernating little brown bats have been known to stop breathing for 48 minutes.

**Bacteria** have been found almost 50 kilometres above the Earth's surface and 7 kilometres below the bed of the ocean.

The African lungfish can live out of water for up to four years.

The feet of a tree frog are very important for its survival. Its feet are like suction cups and this helps it to hold on to leaves, branches and even the trunks of trees. Without these specially adapted feet, mating, sleeping and eating would be extremely difficult!

# Glossary

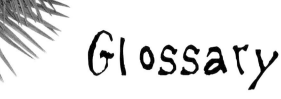

**absorb** to take in (food or water)

**adaptation** changing behaviour or appearance

**antennae** in insects, pair of long sensitive parts that stick out from the head

**bacteria** tiny forms of life that help food to rot, but can also cause disease

**blood vessels** veins carrying blood around the body

**blowhole** single nostril of a whale or dolphin

**canopy** layer of leaves in the tree-tops that forms a roof over the forest

**caustic** burning (like acid)

**dorsal** on the back (of an animal)

**drought** long period without rain

**environment** surroundings or place in the natural world

**evaporate** change from liquid to gas

**expel** get rid of something unwanted

**fertilize** (of soil) make it good for growing things

**filter** prevent unwanted things from passing through

**germinate** begin to develop and grow

**gills** body parts used (by fish) for breathing underwater

**grubs** insect larvae (young form)

**hibernate** to sleep through the winter

**incubate** keep eggs warm by sitting on them

**mate** when a male and female come together to produce young

**mineral** substance that provides animals and plants with extra nourishment

**organism** living thing

**parasol** umbrella that shades you from the sun

**predator** animal that hunts and kills other living creatures

**pressure** downward force

**rodent** animal, like a rat or mouse, with strong front teeth for biting

**roost** to shelter while asleep

**submerged** below the surface of the water

**surface area** total area on the surface (the outside) of something

**tentacles** long thin parts that stick out from an animal's body

# Index

# Titles in the *Amazing Nature* series include:

Hardback          0 431 16652 8

Hardback          0 431 16650 1

Hardback          0 431 16651 X

Hardback          0 431 16662 5

Hardback          0 431 16660 9

Hardback          0 431 16653 6

Hardback          0 431 16661 7

Hardback          0 431 16663 3

Find out about the other titles in this series on our website www.heinemann.co.uk/library